Della Porta

IO. BAPTISTA,
PORTA.

Giovanni Battista della Porta.
Line engraving by N. de Larmessin, 1682.
Wellcome Collection.

The Book of Distillation

Book One

Giambattista della Porta

Translated

by Adam McLean

Alchemy Web Bookshop
2 Craighouse Square
Kilbirnie
KA25 7AF
U.K.

www.alchemywebsite.com

Of Distillation

The First Book

The early appearance of distillation, its causes, and instruments.

Chapter I.

Whether distillation be of the ancients, or of a more recent discovery.

When it was diligently inquired of by curious minds whether the art of distilling in glass was of the ancients, or of a more recent discovery, I resolved to bring it into the middle, which seems to agree with reason. When among the more ancient Greeks I saw no mention of these arts, I judged that they were completely unknown. For Dioscorides, describing the method of distilling pitch, says thus 'It is extracted, while the pitch is being cooked, with clean skins or hides spread over its vapours, and when it has matured, pressed into a vessel, this is done while it is being cooked.'

By another artifice, too, they extracted pure water from corrupted water. On a large pot they arranged wooden bowls, clothed in gauze, or stretched out well-washed wool, immediately placing under them red-hot glowing charcoals, and the vapours were expressed towards the receivers.

Grapaldus received the Arab from Fumanello and reported what he had found. Avicenna, who lived six hundred years after that, in speaking of the phlegm, mentions the alembic and distillation.

'When the superfluities of the food are not excreted in the stomach, they turn into vapours, which, rushing into the brain, fill the head, and turn these by their hardness into moisture, which flows down through the nostrils, as if in an alembic, and flow down under an aqueduct.'

He also mentions rose water. The Arabs, whom I have written about the subject of surgery, say that he was born and educated among them.

Rases and the Albucasi of the same family teach how to extract scented water from roses in the manner in which kings are accustomed to extract it. Geber wrote very clearly about this matter. But I will explain what I think. I have always considered the art of distillation to be true of Alchemy, and born of the same birth, since most of the operations of Alchemy cannot be done without distillation, and while the artisans were searching for the metallic art, they discovered many secrets of distillation. Hermes, the most ancient writer, spoke of the philosophers' stone.

'He carried the wind in his belly', and as all the interpreters freely feel, they say that he spoke of distillation, for the thing dissolved in the breath, seeking the highest and thence falling back, seems to carry the wind in his belly. The most learned Arab philosophers followed him, who wrote about the stone, which was very ancient, but most recently it trickled down to the Greeks and Latins, with whom it lay buried for many ages.

Nicander, speaking of the rose, describes the water, the alembic, and the distiller.

Chapter II.
What is distillation?

Distillation was named by the Greeks, and properly it seems to mean 'to flow gradually', and because the spouts of the stills drip gradually and drop by drop, they therefore are said to distill.

Grapaldus calls the vessel a distillatorium, in which, supposing it to be fire, the simple things are turned into water and passed through the epistomium[1], being distilled. Geber, in his *Summa perfectionis*, says that distillation is the elevation of aqueous vapours in a vessel. But this definition seems to me wanting and empty. For among the species of distillation he enumerates that which is effected by a filter, which is not effected by the elevation of vapours, nor by means of a vessel, nor does distillation take place by the descent by the elevation of the vapours. Some have added, as it were, the power of fire; but rather they should have said of the power of heat, for not only the force of fire, but dung, the slaking of lime, of the sun, and other hot things, generates vapours. Langius thus defines it.

Distillation is the filtering of water resolved by heat into vapour, then condensed again by cold. But in the distillation the moisture is not filtered out: but it is dissolved into vapour, and the vapour is not always condensed by the cold, for in the distillation of strong waters the vapours seeking the head of the vessel do not meet with the cold, but with the very hot, and very fiery, therefore the vapors are not always forced into water by

---

[1] The cock or opening of a water pipe.

3

the cold.

Fumanellus indeed says that it is the extraction (or secretion) of moisture by the force of heat, drop by drop, flowing through a vessel for the preparation of various medicines, following Mesues, who writes that, in simples there are heterogeneous congealed parts, which are separated by the aid of chemists, by the working of heat. Such distillations, or sublimations are neither water nor oil, but a different substance, before you propose to distill them, whence they call the art of distilling and sublimation, as if all distillation were done by sublimation, for generally it is done by descent. But to us this seems more correct.

Distillation is the dissolution of moist parts into vapour by the force of heat, which are turned into liquid by cold and constipation. For the steam is released in the head of the vessel, and if it meets the cold, unless it multiplies there, and is first congealed, it is not turned into a liquid, as we said of strong waters, and when very often the vessels are broken from the greatest constipation, the whole house is scarcely capable of escaping the smoke, which was forced in such a narrow place, as we treated more widely in our *Meteores*, when we discussed the rising of the rains.

Chapter III.

That the art of distilling
has been revealed to us by Nature.

The art of distillation has been demonstrated by Nature in so many ways in all things, that it is a great wonder why it became so unknown to us. Those who saw in the great world, the Macrocosm, were able to remind us that the surface of the earth, heated by the power of the sun, exhaled through the latent passages of its body certain vapours, or rapid spirits dissolved in the winds, which, having been subjected to the ether, meeting the region of the cold sky, meeting again in another form, became streams by their hardness, and sprinkled the earth again with drops. In Man the Microcosm, Hippocrates, describing the sweats in his book *On Flatulence*, says, this indicates distillation.

Whatever the fiery force touches it melts, and these spirits, when they have rushed to the passages of the body, become sweats, since the compressed spirit is transformed into water, and, penetrating through the passages, burst forth.

In exactly the same way that steam, when it has been dissolved in boiling water, if it has an obstacle on which it must impinge, becomes fat and condenses, and the drops fall from those bodies on which that steam has been impacted.

Moreover, we have seen above from Avicenna in his *Catharro*, how the vapours diffused from the body by natural heat to the brain, are forced by their coldness into water, and seek refuge through the nasal canals, and making outflows are expressed as if through the mouth of a still.

Moreover, they saw that when men covered hot dishes in plates, so that they would not cool down after a little while, when they uncovered the plates, considering the moist steam expelled by the heat of the thing was turned into water, they asked for a lid, which, by its coldness, forcing the steam into bubbles, soon bursting into drops it caused them to flow down, so that the subject faithfully was filled drop by drop with water. Our ancestors, taught by these examples, could have mastered the art of distillation more quickly.

Chapter IV.

Of the kinds of distillations, and of that which is effected by ascent through fire.

There are three kinds of distillations, as in motions, distinctions, one extending straight upwards, the other tending downwards, the other slightly depressed or reclining on a slope, and these from the particular propensity of the nature of herbs. For some, of a finer spirit and a finer concentration, are carried straight up, others of earthy matter, the constant, and the feculent, are drawn below, others, having themselves in the middle, desire to be offered on the incline. Therefore, the organs serving the distillation will be compared in a similar way, viz. those that are flatulent and vaporous are distilled, with vessels raised upwards, those that experience spirits and vapours, downwards, and those that lie in the middle are adapted, viz. the spirits of herbs.

Each of these distinctions is again divided into four according to the number of elements, viz., by fire, by earth, by air, and by water, from which it is concluded that there are twelve distinctions of distillation. We will speak about each one, and first about that which is done by fire. We especially use this kind of distillation, when we wish to wring out of the waters the fluids, or spirits, which reside more deeply in the earthy parts, by the great force of urgent fire, namely, to remove oils, or strong waters, so that there is nothing between the vessel and the fire.

I come now to the construction of the furnace. Walls are erected of earth, round or square in shape, four inches wide, two feet in diameter, and a foot in height, built of lime or of strong clay, so that they may not cause injury and suffer damage from the fire.

They should soon be covered over, and above the chamber, a round or square hole should be opened, which can receive a grate of iron rods woven into a lattice, or a floor should be introduced into a trough with many holes, by which the descent of the ashes to the lower part would be allowed. Let a window be opened below, from which the fallen ashes are removed. Above the chamber the walls are raised to a height of half a foot, and immediately above the grate a window is opened about a foot wide, through which the coals and wood are thrown. On the top of the walls you will fix two iron barrels at a distance from each other, to hold the vessel that contains the distillate.

Above the iron rods of the height of one foot are erected vaulted walls, so that above, a mouth is open, through which the vessel is inserted, so that when the fire is exposed, the flickering flames reflected from the dome make a repulsion above the vessel, and by the force of the heat, the vapours rise, which are forced into the head, and from several that soon descend, the drops collecting themselves in the cavity, are immediately expelled into the receptacle.

In the four corners of the tower four round vents, two inches wide, are to be cut out, through which the smoke may escape.

Its form is this.

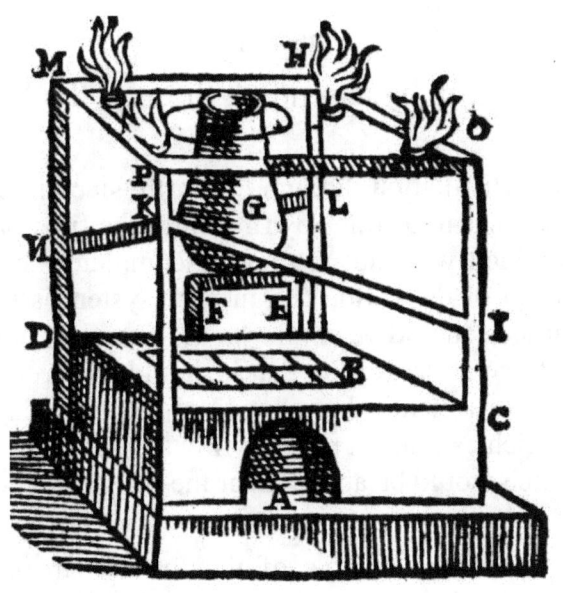

H, M, C, D. Furnace.
A. Lower opening, from which to remove the ashes.
B. Grid.
E, F. Window, from which the wood is thrown.
G. Vessel.
I, K, L, N. Bars supporting an iron vessel.
H, M, O, P. Four vents.

Chapter V.
Distillation by ascending through the Earth.

We use distillation through the earth, since it is milder than what has to be done by fire, since the finer parts and spirits which we want are not so completely hidden in the bosom of the earth. The furnace system is like this. The furnace should be similar to the first one, but with a round dome at the top, so that it should receive a dish in the form of a hemisphere, made of the best clay, or of brass which can take a rougher fire, supported by an iron rod, which would be able to bear the weight of the things placed upon it.

The top of the vessels of the furnace are carefully leveled, and closed with clay, so that the flame or smoke may not escape and injure the outside of the vessel, leaving in the sides, four chimneys, as usual, through which the smoke and flame may escape. The ashes are sifted through a narrow sieve, and spread over the bottom of the hemisphere, a finger high, and placed over a glass vessel, and the ashes are sunk deep until the glass ampoule is covered up to the neck.

But if you wish to attack the matter more vigorously, sift the sand, and with the sharpest iron filings, crush the upper head, so that it receives the neck of the lower vessel within itself, and seal the seams with clay, and allow it to dry, lest it rise up, and spew forth a spirited vapour through the cracks, for you have wasted time and work. Lastly, let the fire be kindled, and let the industrious artificer press on as long as the necessity of distilling the thing required.

Behold the structure of the furnace.

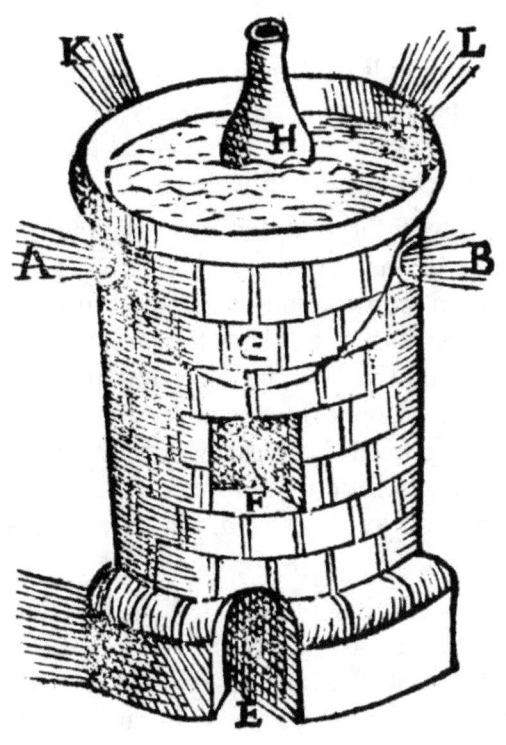

A, B, C, D. Furnace.
A, B, G. Iron rod supporting the hemisphere.
H. Ampoule.
I. Ash.
F. Window for extracting ashes, smokehouse.
A, B, K, L. Smoke holes.

Chapter VI.

Distillation by ascent through the air.

There are two kinds of furnaces which give off hot air, for the heat raised in the air is either by fire or by water. We use the latter when it is more relaxed than when it is to be acted upon by means of heated air.

Let us begin with the first. From the beginning we will form the base of the furnace, which is raised two feet from the ground, constructed of tiles or bricks, so that it does not collapse easily, but lasts longer and is sufficient to bear the weight of the furnace and the vessels. A window should be open in the lower part, from which the ashes falling from the burnt wood or coals may be removed, and a small floor is drawn over it, which in the middle receives an iron covering, which will support the fire.

A wall six feet high should be raised, and this will be the base. Above the base, a tortoise is erected in the form of an egg, the width and height of which is capable of placing the vessels, four feet in width and six in height, so that it can hold twenty or thirty distilling vessels, the wall of this is pierced, in the hollows of which clay vessels or glass in the form of urinals are let in, and as if placed in their beds, they are placed alternately above and below, with such a distance between them that each has its own void below, in which the receptacles are set up to receive the mouth of the upper vessel into which it lies, so that when the distilled water flows out, it is received.

The joints of the vessels with the wall of the furnace

should be blocked up with mud, so that heat or smoke may not pass, but inside the vessels should be reflected. Four air holes should be opened above the base, which transmit the smoke to the outside, and vents should be given to the fire, lest the fire, being obstructed on all sides, should become suffocated.

If the vessels are of glass, the bottom should be covered with yellow plaster, lest they should be broken by the force of the heat, and the labour should be undertaken in vain, and the cost be lost. Let them be filled with simple vessels, and covered with a head, and the side adjacent to each other, be fortified with strawed mud. Let them put charcoals of oak, or of oak wood, on the grates so that they will last longer, and let them set a uniform fire, smoke, and heat the air by asking for maximum heat. And the task of the attendants, who are seated, shall take care that a fire, more violent than that which has been laid out, does not spread.

The furnace will be built like this.

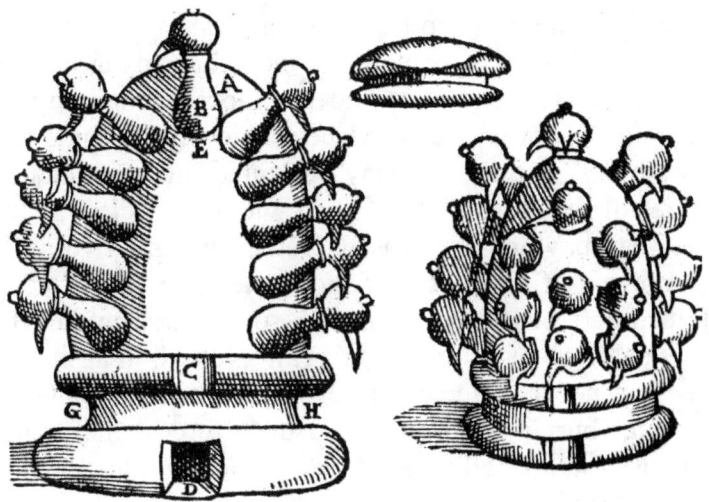

13

A. Furnace part.
B. Distillation vessels.
C. Fire place.
D. Whence the ashes are taken out.
E. Furnace tortoise.
G, H. Base of furnace.

But for those which require the most subtle heat, the same furnace with a furnace full of water succeeds in its substituted function. It should be placed above an iron grate, for if the fire is constantly and brightly lit below, soothing spirits are exhaled from the water, and thin vapours are drawn from the simple ones, without the misfortune of scorching, and it will be of great benefit to us to draw out the waters, that they may be swallowed up by the mouth.

But if it is to be distilled in a vessel, the furnace must be built in this manner. The vessel for distilling must contain a copper vessel containing the material in its reservoir, and if the copper is not available, a pot may be substituted for it. The mouth is indeed narrower, a basin of the greatest capacity, and the mouth containing the contents of the vessel shall be welded to the mouth of the vessel, so that they may be united into one, and the edges of the other shall be lined with sticky clay, so that the heat may not be diffused. But let this great vessel be comprised of a furnace, or above the floor supported by four pillars or if it be fixed with some substructures, the fire shall soon burn brightly all around, that it may become hot and fierce. Thus the air which is contained within, burning beyond measure, forces the matter enclosed in a small vessel to fly higher, and to flow into the underlying vessel.

E, F, K, G. Furnace.
B. Container.
A . Containing vessel.
E, F, H, D. Frame supporting vessels.
I. Window.
L . Opening for ashes.

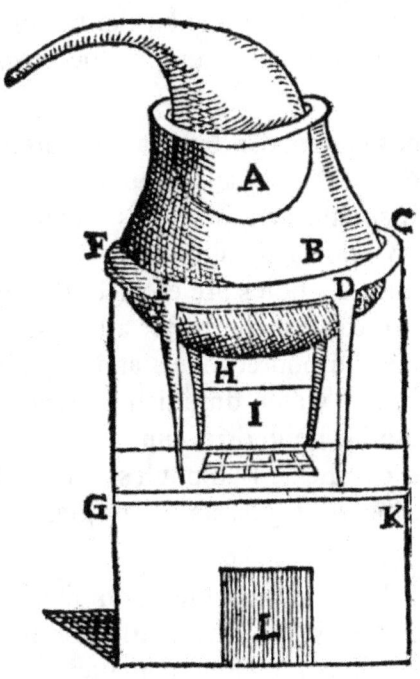

Chapter VII.

Distillation by ascent through water.

Distillation through water, we say through a bath, in which work we use the same as before. In this they differ, that in the upper furnace an empty shell is formed, in this above the bottom of the vessel, grass, wool, or the like, is covered to the thickness of three inches, lest the ampoule should be broken by the harder bottom, and the this is covered on all sides up to the neck of the ampoule.

Above, branches or wooden bars, are secured by some small weight, so that the vessels should not be lifted up by the water, but should lie motionless, depressed by the weight, for they would easily slip out of them unharmed. At the edges of this, air holes are created, through which the fire and smoke can escape. After all these things are done, let the water be poured over, and be subjected to the coals and if the water diminishes by boiling and exhaling, let more be poured in, but warmed, for if the cold touches the overheated ampoules, they will shatter on contact with the cold, just as warm glass is cracked by a cold wind.

A head is placed on the ampoules from that side where their mouths are connected. This distillation is safe from the damage of combustion, for no matter how fiercely the fire is pressed or the coals are burned, the water will never be contaminated by smoke.

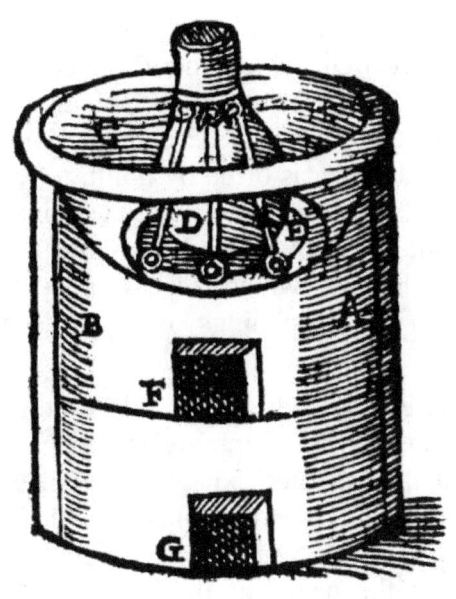

A, B. Furnace.

C. A bronze vessel full of water.

D. Ampoule.

E. A leaden base with which the vessel is secured by chains.

F. Opening for coals.

G. Ashes.

Among us, the artisans use another method of distilling, which, because it does not in any way affect us with smoke, or with the repugnance of smoke in the domestic rooms, let us say that a tower should be made of iron plates, four feet high, and a half foot in diameter, and have two iron partitions with their windows, one in which the fire is stirred up, the other, which should receive the falling ashes. It is supported by three columns below.

At its uppermost mouth it should receive a bronze hemisphere, a foot high, which should be square to the mouth of the furnace; and this will contain water. In particular, the reservoir should be extended outside the canal, so that if the water is dried by boiling, fresh water can be added. The uppermost mouth of the hemisphere is closed by the interception of an iron plate, laced with frequent holes, through which access is open to the smoke, and other large openings of a palm's width to receive the bases of the three ampoules.

This hemisphere is closed with its lid in the form of a lower dome, the edges of which are so received among themselves that they do not exhale the smoke outside, and if it does not work, the four joints are glued together with sticky clay. The tortoise also has three holes in the way, from which the necks of the ampoules pass out. The holes through which the necks are come out are stuffed with cloth or cotton, so that they do not let out the fumes. Let simple ampoules be fitted on the heads with their ampoules well separated, and let the receptacles hang from racks.

Soon the doors of the iron windows are shut against the burning coals, so that they will not exhale any fumes into the room. The water heats up with the urgent fire, the heat penetrating inside the simple vessels heats up, and forces them to fall into their heads, seeking refuge in the cold. The cooling force makes them fall into the beak, and into the hollow receptacle.

As the following picture shows.

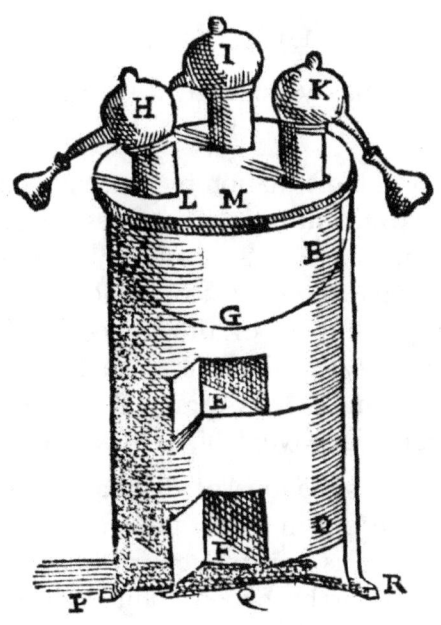

A, B, C, D. Tower of iron plates.

E. A hole for the fire.

F. Hole for the ashes.

A, G, B. The lower hemisphere, is full of water.

L, M. Cover

H, I, K. Three ampoules with their receivers.

N, O. Between the walls inside with its holes.

P, Q. R. Three pillars supporting the tower.

Chapter VIII.
Distillation by descent through fire.

The reason for the distillation by descent, according to Geber, was to extract oil from vegetables, for fats ascend with more difficulty because of their weight, indeed, they easily fall apart below. In order to remove the oil from wood, a great force of fire is necessary, since it sticks more closely in its place, and the wood is inflammable by nature. It is arranged that these things should not become inconvenient, as the oil descends.

There are four distinctions of distillation by descent, each of which differs in form and material, from the others. The furnace is constructed in this way. Let the bricks from the floor be half a cubit high in width to receive the vessel, and let it be sheltered at the bottom. The tortoise is pierced in the middle, except for the neck of the distiller. The floor should be placed above, in the manner of a cavity, which can hold a bellied gourd-shaped flask within it, and the walls above the vessel should be raised by half a foot in its structure to include the flat bottom of the vessel.

Let the distiller take a gourd of earthenware, or coated with cured clay, or of brass, that it may bear the fiery force as long as possible. Fill up to two-thirds of the matter to be distilled with this, and push the narrowed glomerulus into the neck up to its belly with iron beads or auricles. For when it obtains freedom, it will relax itself, and will prevent the material placed upon it from descending.

Cover the vessel, and put the neck through the

opening, from which the neck is placed in the oven, line the opening with clay, and pass the neck into the receiver, and it is clayed, so that nothing of ash or coal may fall into the receiver, which will be accommodated below. The furnace must be windowed at the front, so that whatever the gourd drops from above may be seen to flow into the receptacle. Finally, the fire is kindled around each one, always gradually increasing.

At first you will see the water flowing into the base of the vessel, when there is no material that can contain any moisture; for if you suddenly increase the fire, the excess of moisture is consumed with the oil.

When the water ceases to flow, the oil will descend. Then remove the receptacle and put another to receive the oil; but when the oil begins to flow, increase the fire, so that the vessel burns until it ceases to flow.

On another are set three iron rods, attached above to a circle, like posts, and spread apart in the lower part, to make it more stable like a bench. Within the circle they place an iron shell with a perforated bottom; through which the neck of the vessel is passed, and so they light the fire above.

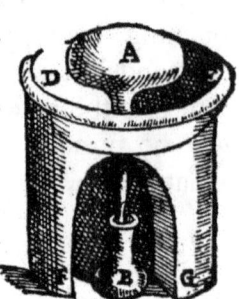

Fornacula,
D, E, F, G,
Cucurbita,
A, Conca
ænea D, E,
Feneftella
inferior C,
Receptacu-
lum B.

E, F, K, G. Furnace.
A. Cucurbite.
D, E. Brass shell.
C. Lower opening.
B. Receiver.

Chapter IX.

Distillation by descent through the earth.

Now let us show the distillation through the earth by descent. We have said that the method of distilling through fire is similar, and through the earth, but by a twofold method: for there are simple things which are less resistant to distillation, and some which are more. Now let us see which suits each one, or neither.

Distillation by descent is done more mildly by the third, therefore we use them, as opposed to the distillation of one, which is more tenacious by fire. Now we will use the present one, since the same stove as above is to be used more gently. There will be a difference between them, that above an ampoule inverted in ashes, two or three inches high, we burn with in a narrow sieve, two or three fingers deep. Immediately set fire to the coals and wood above the stove, so that when it is done, the oil and water will run off below the receptacle.

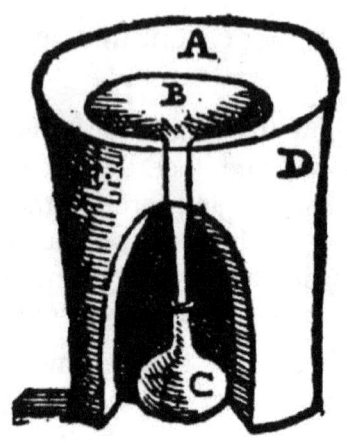

A. The void to be filled with ashes.
B. Distillation vessel.
D, E. Furnace.
C. Receiver.

Chapter X.
Distillation by descent through the air.

Distillation through the air will be a little slower on the way down than when through the ashes. The invention of distillation was to prevent oil or water from being contaminated by smoke, but if the air is heated more violently, the fire rages, and the oil and water are polluted by the combustion. The furnace will be similar to the previous ones, except that above the pillar it will be brass, or equal parts of clay. If it be made of earthenware, it is covered with a layer of lute on the outside, so it resists the fire longer. The lower part of the cavity will be pierced through the middle, through which the neck of the vessel will pass.

The upper part of the lid should project into a round protuberance, whichever suits each one, so that they can be closed very well. An ampoule is placed inside, which has the material to be distilled in its belly, while the neck is passed through an opening below, so the body of the ampoule is half a foot distant from the sides of the vessel. Once the lid is put on, a fire is kindled outside. The heat penetrating through the thickness of the lid, setting fire to the air in the chamber existing in the vessels, sets fire to the ampoule residing in the middle of it. Thus the material will ooze water and oil into the vessel below. When the oil has flowed, the cover is removed with forceps.

But if the substance to be distilled is of a thin consistency, and we fear the weariness of the steam, let the vessel be lightly heated on the outside, and the fire must not remain above it any longer. But when you see that the liquid has been poured out to a sufficient extent, remove the fire.

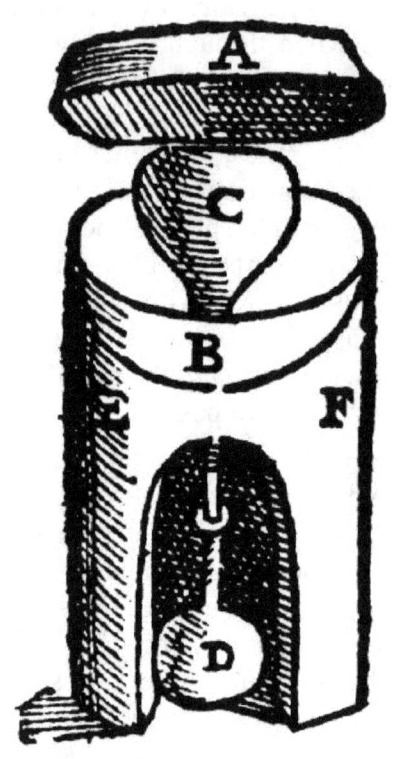

E, F. Furnace.
A, B. Cover of brass bifida.
C. Distillation vessel.
D. Receiver.

Chapter XI.

Distillation by descent through water.

But since we have to deal very gently with things of the thinnest essence, and fear the greatest fumes of smoke, we shall use this kind of distillation. We once devised a vessel for such a work. A furnace rises from the ground three feet high, above the lacuna, a hollow is excavated, so as to receive inside a copper ball, or an egg divided in half. At the bottom it should have a hole open like a brass trumpet, shaped like a funnel, which penetrates to the bottom of the underlying vessel.

After the ampoule is filled with material, and inverted, the neck is put through the tube, water is soon poured in up to the mouth of the tube. A hinged lid is put on, and the loop is tied with iron wire, so that the lid does not fall off when the water swells. The seams of the vessels should be lined with thin mud, so that unless the vents are closed, the swift vapours of water may pass through. A vessel should be placed below, so that the neck of the vessel may be received within it, and the mud should be lined with it, lest a drop of water, which falls on the belly of the vessel, penetrates inside, and disturbs the oil and water with its mixture.

As soon as a fire is kindled above the vessel, which like water heated, will fill the chambers of the vessel with its hot vapours, the spirits below the vessel will be disturbed, which, forced by the harshness of the cold, will go into the water and the oil.

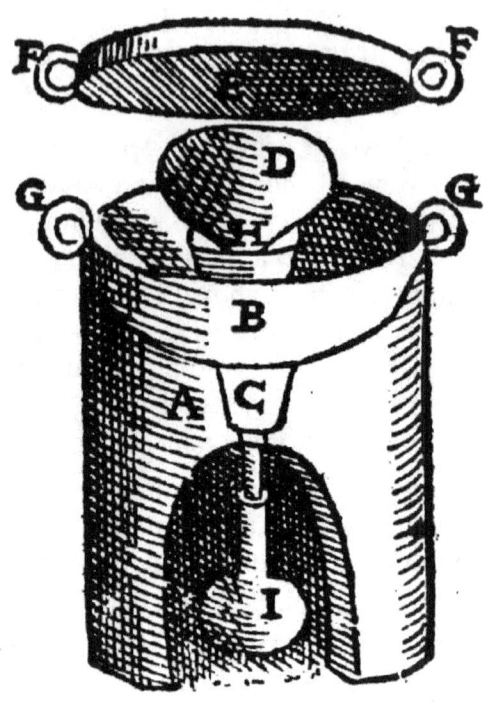

A. Descending furnace.

B. Hollow brass vessel.

G,G. Loops.

C. Glass ampoule. Brass trumpet.

E. Hinged lid.

F, F. Loops.

I. Receiving the subject with the neck turned towards the dripping fluid.

I. Receiving vessel.

Chapter XII.

Distillation by inclination through fire.

We have spoken of the distillations by ascent and descent. Now we tell of those things which are in the middle way, that is, by inclination, that is, by sloping sides. In these we shall use various furnaces, various vessels, and methods of distilling, especially in things which have steep ascents and descents, which is sought in such a way, as for strong waters, and oil of vitriol. A round or square brick furnace is made with two distinct chambers, the upper part of which is shielded, so that the flames flickering through the shield are reflected onto the bottom of the ampoule. The lower part should contain bars of iron, which will be windowed.

An iron rod shall be formed in the middle of the walls, which shall be in the shape of a semicircle in the middle, of such thickness, that both the force of the fire and the ampoule may bear a heavy load. Wood and coals are thrown over the grate, and using iron tongs, the coals and wood are piled up around the flask, and when they have turned to ashes, they will fall down through the grates.

At the four corners are set four chimneys, through which the smoke may pass out. When the furnace is ready, let a glass ampoule be taken, made of the best glass, pure and without spot, which must be coated with sticky clay, so that it can bear the heat and incandescence of the fire. The neck of the furnace is let out through the hole, and it must be closed behind, so that no air is allowed to escape.

A vessel will soon be fitted to the outside, into the mouth of which the neck of the ampoule is inserted, and the joints are covered with clay, lest they emit burning spirits. The vessel should always be kept wet with linen cloth, and the sponge should be kept cool, lest, if the simple distillations were breathy, they might be blown apart by strong breaths, and fall into pieces, and the oil and the works perish.

While the coals are burning, if the bars of the grate are too narrow, so that the ashes do pile up, and exclude the access of air, you must cover them with an iron hook, so that the fire does not suffocate. It is to be taken care, that the fire may burn without interruption, with assistants working alternately, so that those who have completed the day's office may rest at night, and perform the night at another.

Thus, on some days, oil and water will pour out from the burning coals. But when you see the receptacle filled with white smoke, it is a sign of the completion of the work, so do not press the fire any further. Remove the vessel from where the fire resides with the ashes, and collect the oil and water.

A, B, C, D. Furnace.
E, F. Iron supporting a hollow vessel.
G. Ampoule.
S. Tortoise.
H, I. Iron grates
K, L, M, N. Four smoke holes.
H. Opening for coals.
O. Opening for ashes.
P. Receptacle.

Chapter XIII.

Distillation by inclination to the ground.

We will use this distillation, when the simples swell into bubbles with a barely available heat, so as to fill the vessels with their growths, and through the neck of the vessels they are thrown into the receptacle, as they are almost all resinous.

A square or round tower rises from the floor. Let it be a foot high above the ground, and there should be a partition between them, in which the iron bars should be placed at the top, but with the turtle inverted, so that the floor above should sink into the hollow. That cavity should open with a wide hole, in which the belly of the vessel can sit, the neck being extended in the right direction, protected by clay.

The vessel called a turtle, or leutum, is filled with things to be distilled. On top is scattered the ashes separated through a narrow sieve, a finger high. Let the coals, and the water, and the oil be burnt above. When they begin to flow, increase the fire, for the heat will check the growing bubbles. They will dissolve in the funnel, and the oil will flow. Then a light fire must be lit below, so that it may flow more quickly without fear of excrescence, so that the oil may not be polluted by the decay of the burning.

A, B. Round tower.
C. Vessel.
D. Opening for fire.
E. Opening for ashes.
F, G. Fire lit above the vessel.

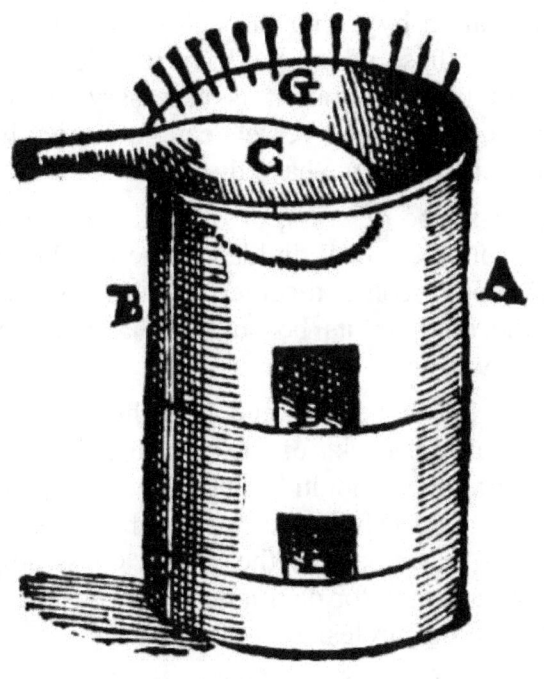

Chapter XIV.

Distillation by tilting through the air.

This distillation by tilting through the air, is not very different from those which we have said above, it only differs from these in that it is necessary to devise a method for how inclined vessels can conveniently distil through the air, which we have discovered.

Let the vessel be of brass, or of hard clay: but of the best earth, round or oval; divided in the middle: but the divisions should come together in turn, so that one receives the other in its bosom, so that they scarcely transmit the vapours.

The lower hemisphere should have holes at the top, through which the necks of the ampoules should pass, the vessels of which should be supported by iron circles, which are suspended from the sides of the vessel, so that they hang in the middle of the vessel. Glass ampoules are filled with things for distillation, and must be placed in the midst of the circles.

The lower half of the hemisphere is filled with water and then covered, and the loops are tied with iron wires, and so that the force of the spirit does not close it, the seams are glued together. The vessel is placed above the furnace described above, and the coals are set on fire downwards, so that the water boils, which with its hot vapours ignites the air, and the air drains the glasses, and the water and oil flow into the underlying receptacles.

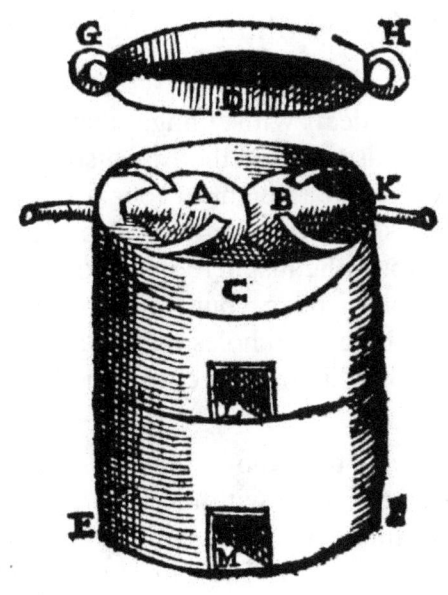

The true vessels for distilling.
C. Lower hemisphere.
D. Cover.
A, B, E, F. Tower.
G, H, I, K Loops.
L, M. Windows.

Chapter XV.

Distillation by tilting through water.

If we have to deal with things that need gently distilled, we will use this method of distillation. When the walls have risen to a height of four feet, as we often see, a reservoir is excavated in the upper dome, which will receive a bronze hemisphere in its round pocket, as we have described above, but in the meantime the difference is that it has two holes on the lower side, with its canals, through which the necks of the ampoules are placed.

The ampoules are tied above the iron plates, so that they do not rise and move with the water, and when the vessel below is filled with water, that the ampoules of the reservoir sink under it. The canals are blocked with cloth and linen, so that they do not emit water and spirits.

After the coals have been subjected, the hot iron, and the boiling water dissolves the simple vapours, which are collected from the cold into the condensed water from the subjected vessels. We have used this vessel in the distillation of the flowers of Rosemary, Jasmine, Violets, and similar powdered flowers.

Chapter XVI.

Distillation by the heat of the sun.

We have completed the distillations by the heat of fire, now for the sun, dung, wine, feces, its substitute for lime, and the like. They use the sun, not only by itself, but by reflection, not only in the absence of fire, but for the greater power of the medicine. We have also seen that the water extracted by fire from the simples has received some unusual qualities and bitterness, which does not come from the sun, as we have experienced in cosmetics for the eyes.

We also use it in scents, which, as much as they fear burning, are so thin in essence. A bench is prepared three feet high, half a foot thick, and of such a length as to correspond to the nature of the vessels to be placed upon it. That part of the bench, which faces the sun, is closed with boards, so that the sun, falling on the receptacles and heating them, forces the water to go away again into steam, and return from whence it came.

Where the sun has entered Gemini (for only in the summer season we will be able to use this method of distillation) a bench under the god of the sun is set facing it.

In the morning, after the herbs have been well washed, and the vessels filled with the dried herbs, the necks are introduced into the mouths of the receivers, and the vapours, when they are made, will be relaxed, and will prevent the exit of the simple.

The necks of the vessels are soon submitted to pass through the holes of the bench. Underneath are the vessels to receive the necks of the ampoules, and soon the vents are closed. For the sun, ascending to the middle of the sky, will heat up the vessels of the flask with the most intense heat, so that the herb dissolves into a liquid, and afterwards into a powder, which itself flows drop by drop into the lower vessels.

A, B. Bench.
C, D. Vessels.
E, F. Holes.
G, H. Receiving ampoules.
I, K. Table for holding objects in the sun.
L. Shining Sun.

But in cold regions, where the oblique Sun does not heat up so much, another method will have to be used, namely by the reflection from mirrors. For a concave or better, a parabolic mirror, cover the sun which turns the reflected rays to the vial, for the vial to be distilled will stand between the sun and the mirror, for the rays, reverberating into the reservoir of the ampoule, will not only produce heat, but will arouse fire, as more widely explained in our book on *Natural Magic*.

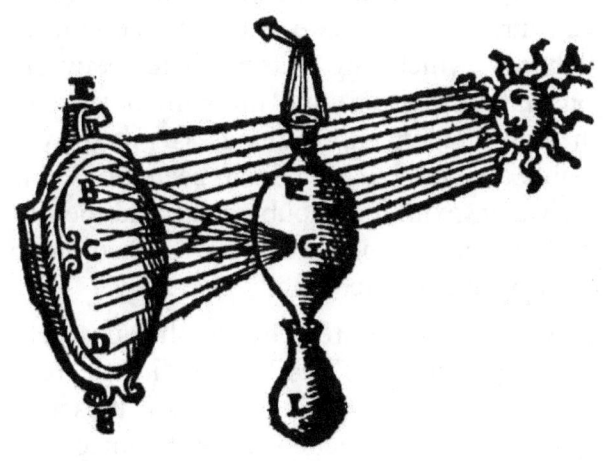

A. Sun.
A, B, C, D. Rays.
E, F. Mirror.
B, G, C, G, D, G, M, G. Reflected rays
L. Ampoule.
I. Receiver.

Chapter XVII.

Of distillations by other heats.

There are also other kinds of heat, such as grapeskins, the dregs of olives, which are left after the expression of oil, be it horse, or beef, and quicklime; which can also be called putrefaction. There are also certain kinds of distillations, which are intermediate between putrefaction and distillation, and this will be when a shell, which will contain horse-dung, is fitted with holes accessible above the bath, which, heated by a furnace of boiling water, exhibits a double heat, and the distillation is hot and moist, and is a sort of medium between putrefaction and a water bath.

Grapeskins, and the dregs of the olives, which are left after the expression of the oil, is the first degree of heat, and we use the softest things, we use those which rot with a little heat. The second degree is horse dung, or ox dung, the third and worst degree is quicklime, steeped in water, and the heat is hot and dry. Those who wish to use dung and hot water should use this furnace. Let it be a wooden box six or ten feet wide and four feet high, through the middle of which a bronze channel runs, the thickness of a human arm, perforated, so as to be rife with fissures and holes on all sides.

Let the box be filled with horse or ox-dung, containing within it glass ampoules, packed on each side with their caps, and receptacles for the things to be distilled. A channel should be brought out, the mouth of which should be fixed to a vessel full of water with its lid well closed. The vessel shall be adapted to this furnace, by which, being heated by the fire, the water in the vessel passes through the canal, and heats the horse's dung through its holes, and causes it to be filled, and by the heat of its own simple things, so that they are distilled in the vessel.

A, B. A similar wooden box.

D, E, C. Distilling vessels.

E, F. Channel with small holes running through the middle.

G. A bronze vessel from which the channel runs.

I, K. Furnace.

H. Opening for the fire.

Chapter XVIII.

Whether the qualities of the simple are extracted in the distillation.

Not many distillers, some envious despisers of the medical arts, and gross ignorants agree among themselves, and are distracted by dangerous opinions; whether by simple distillation they obtain superior or ineffectual qualities, and these go into their own and similar believers, or into the opposite.

It is a matter of the greatest consideration, and of wonderful utility. Manardus says in his *Epistles* that nothing, neither taste nor smell, is obtained by distillation, but mostly passes into various and opposites.Its is

Absinthe, he says, endowed with bitterness, and endowed with much fragrance, gives a water distilled through a lead alembic, which neither absorbs the bitterness, nor smells of absinthe.

Ocymum[2] the most fragrant distilled spirit, not only produces no odour, but is rendered more disagreeable, all of which lends credence to the fact that the water does not flow forth endowed with the same powers with which the plant flourished.

Cardanus has ventured to decide the opposite, and perhaps more correctly, for not only of the same quality: but you can learn by length the greater ones, by heating, moistening, drying, and the like. An example is distilled wine, which produces burning water, which possesses stronger powers at length, and the more it is distilled, the hotter, sharper, and more penetrating it becomes.

---

[2] Basil.

Rose-water is extracted from the flower, of excellent colour, fragrance, and strength. A pitcher of plantain water stops the blood oozing from all sides, and lettuce is endowed with the same powers, if the stronger nature does not possess the same powers. Melissa, which has matured in wine for three days, reclaims memory from oblivion, and therefore, moved by these experiences, Cardanus takes a contrary view.

Distillation makes the water always thinner and warmer. Aristotle says in the *Problems*, that the sun exerting its forces on the waters, draws at the same time the lightest and sweetest parts, which he took from Hippocrates, in his book on *Air, Waters, and Places*, that the rainwaters are the smoothest, thinnest, and sweetest, because the sun draws from the waters that which is thinnest and lightest, and for this reason the sea is salty, because what is heavy, thick, and salty remains, when the sun cannot bring it out.

But if the sun gently warms these things, what can we expect from a fire simmering in stills? and that it is light and thinly hot, because it approaches the nature of fire. But while we were meditating on these things from the various waves of opinions and anguish of mind, we judge these things in this way. First of all, it must be known that every living thing lives by three substances. Firstly, the alimentary, by which each plant is nourished, and daily draws from the earth and from the sky, and when these are lacking, life ceases.

This is more clearly seen in the grasses which do not live on the very skin of the earth, and when it descends to the ground, the sun heats up beyond measure, and dries up more deeply, the roots drying up every day, deficient in moisture. If one sets the green branches of plants on fire, they burn more intensely, because they are abundant with moisture, the food they have drawn from the earth, and if they are set on fire from one end, they exude moisture from the other, which is driven away by the fire.

If one injects the green plants with steam, they immediately dissolve into moisture flowing drop by drop, from dry ones more slowly, sparingly, and perhaps not at all, because they lack food and moisture along with life.

The second is called substantial moisture, in which the substance of plants resides, and here it is not deposited on the surface, but interiorly, and flows down in distillations after the alimentary, and is more difficult to remove: for when it resides partly in matter, partly in moisture, the same distillation is made in turn a third and a fourth time, by pouring it over the dregs, by digestion, and finally by putrefaction, as we shall show in its place.

Fernelius said that this humour is in the white oil, and it is true that it is substantial in red, but I judge that there is little difference between the white and the red oil, and with so much vigour, it seems to me that he has failed in this.

The third is the vital humour, in which the whole substance, virtue, and property resides, and it is in the oil, that the odour, flavour, and virtue is found in the highest degree of excellence, the proper seat of spirit and heat, and in which the whole essence is cherished, and when it is extracted, nothing but the corpse of the simple remains, as a body completely bereft of life.

It will be very difficult to determine the question even with these preliminaries, since the powers, natures, and properties of plants are varied and multifarious: for some are in waters, some in oils, some in the salt itself, and others are hidden in the blind recesses of the bodies, as in this whole work we will see. That others say that strength does not remain in water, citing the example of absinthe water, which is sweet and does not contain any bitterness, which results from lead.

For what else is white lead than a sheet of lead spread over a wide mouthed clay basin? and when it is infused inside with the most acrid vinegar, the breath of which is carried to the top, it dissolves the lead into white lead, from whence it is scraped, or put collected in a place, and it has a sweet taste.

When, therefore, in a leaden alembic, the sharp fumes of absinthe seek the top of the helm, they dissolve the lead into the ceruse or white lead, which, falling down, falls through the channels into the receiver, and appears to be thickening when it is poured out.

Hence it should scarcely seem surprising if it does not recover his bitterness, and he did business with his stomach. How much does he deserve, and does he deceive his own bitterness? And if you wish to become more certain, you should instill a drop of oil of vitriol into the water of the rose, or of the orange blossoms, and immediately you will see that its clearness is disturbed, and that the white lead will settle in the wine.

But if it is distilled in water vapour in a leaden alembic, or in a glass vessel, it will retain its colour, taste, smell, and strength, and will not be infected, and this will be more conspicuous in its oil, for to the highest degree, the smell, taste, and heat will be found in it.

Whence these things cannot be confined by general

precepts, so the properties of all things diverge. For this reason we shall enumerate the qualities of each in our own chapters, lest we imitate those whom we criticize, when we must find them false, comprehending the nature of things by a single rule.

This, however, cannot be affected, but by repeated distillation it receives thinness and heat, and reduces the humidity and coldness. Oats are of cold quality: but the water dripping from them touches the head, and intoxicates. Among the Tartars it is reported that this when distilled is intoxicating.

Chapter XIX.
Of the various kinds of vessels.

As there are so many kinds of things to be distilled, so there are many kinds of vessels, indeed, from the various geniuses of the inventors, that each one through his own will fashions and shapes various things, so that they are found to be almost infinite, oblong, short, curving in on themselves, remaining curved, broken, and spiral. But I proceeded with the greatest ingenuity and skill of the most diligent investigators of the virtues of herbs, and some have been devised so excellent, that without them it may be thought wanting and difficult to some; for the distillations adapted and used the talents, qualities, and natures of the simple distillers, and these were borrowed from the forms and natures of animals.

Where animals, which are the thinnest spirits, breathe high up through a long windpipe; the spirit of a plant is very thin, and it is easily lifted to the top, for this reason vessels with a longer neck are used, i.e. they are sublimated for a longer path, the length of the neck being eight feet, which is marked by the letter A, shown below, which we use most frequently in distilling the water of life, for through the longest journey the spirit ascends through the humidities and is relieved of its dross, and emptied of its contents, so that by working it acquires the greatest thinness. The common people call it matarozzo, which we call the ostrich, giraffe, or gruale vessel.

On the other hand, if things have simple terrestrial dryness, and are thick and at least partially vaporous, and flatulent, for their distillation large and equally humble organs are required. The tortoise is an animal quite terrestrial, dry and covered with a rigid covering made of earth, much of which it consists, with a large body, always prone, and walks with its head bowed, homelike, slow-moving, like simple terrestrials. We use these tortoises for salts, vitriol, and other minerals, which do not ascend. The common people call a lute, because it is shaped in the form of a tortoise, or the musical instrument called a tortoise, or a lute.

These vessels, lying like a shell, as a tortoise also walks, are placed in the furnace, so that the liquid, drawn up by the force of the fire, immediately flows into the underlying flask.

....

They are also simple moist parts, thick but with little vapour, so that they imitate a bear, thick, earthy, slimy, stupid, shapeless, so that the whole body and head appear without a neck, with a large, fleshy, short body. They form a vessel in its likeness, which the common people call an ursal, or a urinal, in order to remove the viscosity and erosivity of simple, less spiritual, or difficult ascents, and carry them over in a short, wide tract.

But if the simple parts of the spiritual essence are immersed in coarseness and earthly dregs, so that they may escape thinner and purer, and the coarser and impure ones are freed from them, and left in the dregs, as if unfit for many uses, it is necessary that, by multiplying their forces, they should be resolved within themselves, and brought back, so that surrounded by constant movement they may be allotted a nobler force and a more excellent one.

A vessel was devised, which they call a Pelican, which was formed in the shape of a Pelican bird; in which the more slender parts of the simples, brought out through the neck, and thrust through the beaks back into their open chest, are implanted and carried back through the air onto the dregs, and again lifted up through the neck. By this untiring movement, they gradually drain away the wateriness, the fatness, and by simple constant rotation they are not only purified, but the virtues are even more highly exalted.

The vessel is marked with the letter E.

They portray others in a different way. They take two vessels, each being connected to the other, and what the one receives gives back to the other. Each is fastened to the belly of the other by the beak of the other, just as twin brothers embrace each other, hence in the honorable nomenclature they call these, twin vessels, that they may be thinned out by frequent circumambulation, and from all filth and dregs purged, they may acquire nobler powers.

There are also some simple ones, which are forced to be carried along a long and winding path in order to acquire a more sublime nature. For while the Serpents ascend higher through these windings, they continually leave the moister and heavier parts, which are detained below, while the purer ones are advanced higher, and because it more favorably imitates the windings of a serpent, they are called serpentine.

We, too, have often used this in the extraction of the water of life, so that a finer, purer, and more penetrating nature may be drawn into the sublime through narrower winding paths, and production and things may be freed from all phlegm.

F. Vessel.

To bring out the water of life we also use a certain vessel, which consists of eight or ten heads, the top of one of which is mutually inserted into the body of the other.

While the force of fire is raised with the greatest urgency above the living water, it completes the water of life of various qualities, and the higher it is raised, the thinner it is, and moister, thicker, and overflowing with phlegm, the more ignoble, because the necks and heads of hydra seem to be imitated, we call this the seven-headed hydra of Hercules, but in one case they are made of bronze, or of auiricalcum, and in another of glass. The vessel is marked with the letter G.

We also use a curved vessel with a neck: but with a large channel in which those simples can at least ascend. The common people call these storks, and we use them in distilling strong waters. And because the neck twists, we call it twisted, so that those things which defy sublimation and are persistent, by the force of an urgent fire, are drawn more willingly through the twisted neck.

Nor have we refrained from naming vessels after the similitude of plants and fruits, for in distilling simple things, which have strong vapours, and pressing spirits into salmoniac, and salt water in strong waters, we need a large-bellied and capacious vessel, and because it is shaped in the likeness of a gourd, we call it a gourd: for if we try to control that force of spirits in a narrow vessel, they go off with a loud bang, and the vessels scatter into a thousand pieces. It is necessary, therefore, to have a large vessel, in which those violent spirits can roam more freely; so that by the cold that occurs, flatulent vapours may be dissolved into a liquid.

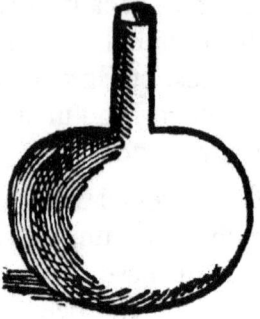

Chapter XX.
The matter of distillations.

The ancients, while the art of distilling was still crude and imperfect, used lead or copper vessels in extracting the essences, which were so destructive to human bodies that they fell into various strangely noxious diseases. Galen, in his *Book of Medicines*, the second part, advises that water which is brought through leaden canals should be regarded with horror, for they retain a kind of slime and turbidity, so that those who drink such water in abundance are subject to dysentery.

Apart from the smoke, and the fungus, they do not waste their time with marked disgust; but for those who drink the water, it is not only troublesome, but in a strange way noxious: for it affects the stomach, the chest, the liver, and all the viscera. And if any one has the heart to try it more clearly, let him observe water that has been sprinkled with rose or orange blossoms, and if it has settled for several days, he will find sitting at the bottom a white lead (as we have said) which was made from the lead alembic. However, so as not to be tainted by the fact that those who use a lead alembic draw out more water, since by its coldness, the rising vapours are forced into drops, and thence it exudes more copiously through the orifice.

Nor is it unlike copper or iron vessels, for they contract verdigris or rust from the distillation of simple acids, which, when mixed with water, brings danger to those who only absorb it, as much as the verdigris and rust itself are more harmful than the white lead.

But better than these are earthenware vessels, glazed inside, and those who fear them being damaged by the fire, will be able to make the clay with straw so that it will last longer in the fire.

But the best of all are the glasses, for the purest glass, receives no foreign impure quality, and retains the powers to which it is given carefully.

But metallic vessels are to be used in extracting oils, where the innate power resides, at least for chemical purposes, or for coating medicines applied to the body, but not to be absorbed by drinking.

Chapter XXI.

Of the various degrees of fire.

It has been consulted among the distillers to distinguish the grades of the fire into degrees, so that the appropriate degree of heat may be established for each one; for they are simple, some very moist and light, some resistant to cooking, so that they demand a more vigorous fire. It is the opinion of simple people that the nature of fire should not be violently increased; the opinion of the philosophers is that Nature should not be forced too violently, nor is it allowed to be violated, for it is corrupted by too much coercion.

The first degree of fire is the bath, when the water is hot enough to put the hands into it without injury; from this fire, flowers, and fruits, and similar soft slimy substances, or phlegmatic, thin, and evaporable, which generate a light vapour with every heat.

The second stage, when the water is very hot; yet it does not bubble up, nor can it be touched with the hand, and it is the fire of simple, thicker substances, which, owing to the coarseness of the matter, are not so easily vaporized.

The third stage, is when the water boils.

Fourthly, when there is nothing between the fire and the vessel, which we use in distilling strong waters.

There are those who also divide the degrees into three parts, so that the second part of the first degree is the temperature of the water, and so on for each.

Alchemical Translations Series

1. Monte Raphaim - The Morning Redness
2. A Spagyric and Philosophic Revelation
3. Allegory of the Lesser Countryman
4. The True Practice of Nature
5. Chymical Moonshine
6. Pordage - Philosophical Epistle
7. The Great Work - Grillot de Givry
8. The Fountain of the Wise
9. The Fountain of Bernard revealed
10. The Masonic Philosophical Cross and Cubic Stone
11. The Flower of Flowers and the Path of Paths
12. 134 Woodcuts of chemical and alchemical apparatus
13. Dicta Alani and the Mirror of Alchemy
14. A hundred and twelve accounts of transmutations
15. The Garden of Riches
16. The Strange Guest - Gustav Meyrink
17. Sunflower of the Wise and Four curious letters
18. Webster - The Transmutation of Metals
19. The Mirror of Philosophy
20. The Allegory of John of the Fountain
21. Arcanum arcanorum arcanissimum
22. Heavenly Manna - Azoth and Fire
23. The Magic Cave in Scotland
24. An Alchemical Reading of the Song of Solomon
25. An Allegorical Alchemical Journey to the East
26. Dialogue between Chrysophilus and Theophrastus
27. Philosophical Axiomata by George Ripley
28. The Twelve Grades of Alchemy by J.D. Mylius
29. The Birthing Bed of the Philosophers' Stone - Nollius
30. Discourse on the Philosophers' Stone
31. Two French Alchemical Allegories
32. The Allegorical Discourse of Solinus Saltzthal
33. The Book of the 22 Hermetic Leaves
34. Alchemical Visions and Allegories

35. First Book of Distillation - Della Porta
36. The Philosophical Parergon - Nollius
37. War of the Knights - Limojon
38. Dialogue - Aegidius de Vadis
39. Donum Dei - Samuel Baruch
40. Banquet of the Sages
41. Transformation of the Metals - Denis Zachaire
42. Allegory - Eirenaeus Philalethes
43. The Memorial of Alchemy - Pierre Vicot
44. A Philosopher and a Peasant discuss Alchemy
45. Transmutatory Alchemy - Timothy Willis
46. Light out of Chaos - Louis Grassot
47. The Play of Children and the Work of Women
48. The Secret - Jodocus Grever
49. The Metamorphosis of the planets - Monte-Snyders
50. A Philosophical Riddle - Birkholz / Adamah Booz
51. The guide to the chemical heaven - Jacob Toll
52. Mercury's Caducean Rod - William Yworth
53. Centrum Naturae Concentratum - Ali Puli
54. The Fate of the Alchemists
55. A Cabalistic Fable - Monte Hermetis
56. The Philosophical Bird-Catcher
57. Truth of the Philosophers' Stone asserted
58. Chrysopoiea
59. Twelve Royal Palaces of Hermetic Wisdom - Fictuld
60. The Mystical Cabbala of Nature - Fictuld
61. The Pilot of the Living Wave
62. Philosophia maturata
63. Chaos - Fictuld
64. The Open Ark
65. Steganographic Allegory - Beroalde de Verville
66. The Aphorisms of Geber
67. The Secret Fire - Glauber
68. Anthroposophia Theomagica - Vaughan
69. Allegorical Dream - Fictuld
70. More - Ezekiel's Vision of the Mercava

71. Chymist's Key - Nollius
72. The Fama Mystica
73. The Crowning of Nature engravings
74. Cabala Verior
75. Zosimos
76. Three short alchemical texts
77. The Three fires of the Sophi
78. Rosicrucian Preface - Sperber
79. Of the Elements and the Quintessence - Drebbel
80. Treatise on the Philosophers' Egg - Bernard of Treviso
81. Ariadne's Thread
82. The Philosophers' Stone - Vauquelin des Yveteaux
83. The Inferior Astronomy
84. The Tomb of Poverty - Henri d'Atremont
85. Compendium Hermeticum
86. The Great Work Unveiled
87. Tinctures of the Seven Metals - Basil Valentine
88. Letter from Hephaestion to Alexander the Great
89. The True Hermes
90. Nature Uncovered
91. The Chemical Truthsayer
92. The Blood of Nature - Brummet
93. An apologetic treatise - Robert Fludd
94. The Key to the Great Work - Sancelrian Tourangeau
95. The Mytho-Physico-Cabalo-Hermetic Concordance
96. The Reign of Saturn - Huginis a Barma
97. Salt, Light and the Spirit of the World - Nuysement
98. A Philosophical Letter - Philovite
99. Writings of Cleopatra the Alchemist and Maria Prophetessa
100. A Rosicrucian Colloquium
101. The Golden Treatise of Xamolxides
102. The Golden Rose
103. Explanation of the Emerald Tablet of Hermes - Garland
104. Mercury Revived - Samuel Norton
105. The Compass of the Wise - Birkholz
106. The Silence after the Clamour - Michael Maier

107. The Golden Mirror of Outer and Inner Vision
108. A Hermetical Banquet
109. Addresses to the Gold- and Rosy Crucians - Ecker
110. Chrysopoeia - Augurel
111. The True and Perfect Preparation - Samuel Richter
112. The Hellish Goddess Proserpina - Rudolph Glauber
113. The Powder of Projection - D.L.B. Lord of la Borde
114. Gold Unclothed - Johann Christian Orschall
115. The Divine Arcana
116. Three Curious Alchemical Writings
117. Theory and Practice of the Gold and Silver Trees
118. The Philosophical Water
119. A Treatise on Metals and Alchemy - Bernard Palissy
120. Chemical Essays - Karl von Eckartshausen
121. The Aurora - Henri de Lintaut
122. Of Hyle, the Universal Prima Materia - Khunrath
123. The Four Amphitheatre Engravings - Khunrath
124. The Fire of the Magi - Khunrath
125. The princely and monarchical Roses of Jericho - Fictuld
126. The Oraculum Manuscript
127. Kings of Scheschian
128. The Theoricus or Second Degree of the Rosicrucians
129. Nine treatises on Goldmaking - Stephanos
130. The Philosophers' Stone - Athanasius Kircher
131. Aelia Laelia Crispis - Nicolas Barnaud
132. Sphynx Rosacea - Christophorus Nigrinus
133. The Theory of the Divine Art of Alchemy - Mylius
134. Summum Bonum - Robert Fludd
135. The Philorcium of George Ripley
136. The Hieroglyphics of the Egyptians - Michael Maier
137. Hermes - Emblems of the Twelve Nations - Michael Maier
138. Maria the Jewess - Emblems of the Twelve Nations - Maier
139. Light emerging by itself from the darkness
140. The portable laboratory - Johann Joachim Becher
141. The Hieroglyphics of the Greeks - Michael Maier
142. On the Creatures of the Aethereal Heaven - Robert Fludd

143. On the Beginnings of the Macrocosm - Robert Fludd
144. On the Fabric of the Macrocosm - Robert Fludd
145. The Philosophical Pleiades
146. The Philosophers' Stone - Hildebrandt
147. On the Philosophers' Stone - Libavius
148. Testament of the Fraternity of the Rose and Golden Cross
149. On the philosophical works of the Stone of the Sages - E.H.
150. Text of Alchemy and the Green Deam - Trevisan
151. The Abandoned or Lost Word - Trevisan
152. Secret Manuscript of the Fraternity of the Rosy Cross
153. The Work of Peter of Silento
154. Two alchemical conversations
155. Golden Age Restored - Hinricus Madathanus
156. Monsieur Dupuits - An Explanation of Flamel's Figure
157. Commentary on Sendivogius - Orthelius
158. Three allegorical alchemical texts
159. The Fortress of Science - Grick
160. The Work of the Green Lion - Jacques Tesson
170. Two 15th Century alchemical works
171. The Adventures of the Unknown Philosopher
172. Rituals of the Rosicrucian higher degrees
173. The Alchemy of Flamel - Denis Molinier
174. Olympus Explained - Jean Vauquelin de Yveteaux
175. The Labours of Hercules - Michael Maier
175. Introduction to alchemy - Georg Wolfgang Wedel
176. The Philosophical Master Key - Franz Clinge
177. Urim and Thummim - D Mensenriet
178. Zoroaster: The Key to the Art
179. The Perilous Fountain
180. The Most Secret Mysteries of the High Degrees
     of Masonry Revealed

www.ingramcontent.com/pod-product-compliance
Lightning Source LLC
Chambersburg PA
CBHW070821220526
45466CB00002B/731